Printed in Great Britain
by Amazon

34377498R00069